Keto Meals For Everyone

Delicious keto meals for revitalize your body and weight loss

**Celine King**

TABLE OF CONTENT

Introduction

The ketogenic diet is a very low carb, high-fat, and moderate protein diet. The idea behind the ketogenic diet is that the liver will break down fatty acids into ketones, which can fuel the brain. Ketogenic diets are often prescribed for children with intractable epilepsy because they may experience seizure control while on this type of eating plan. There are many reasons to try a ketogenic diet, but the biggest reason is weight loss. This diet allows you to lose weight without training a lot, as your body burns fat for energy instead of carbohydrates. As a result, you can have more energy, feel better, and have fewer cravings.

An essential step in any diet is considering and establishing goals—thinking hard about being able to precisely articulate what you wish to achieve by getting started on a diet. Because of the temptations of food and treats that can cause us to watch out for health-wise, a keto diet is a fantastic option due to all of the ailments it can help prevent or supplement.

Most people can safely seek out the keto diet. Nonetheless, it is best to talk to a dietitian or doctor about any significant diet changes. It is the case for those with disabilities underlying it.

A successful treatment for people with drug-resistant epilepsy could be the keto diet.

While the diet can be ideal for people of any age, children, and people over the age of 50, infants may enjoy the most significant benefits as they can easily adhere to the diet.

Adolescents and adults, such as the modified Atkins diet or the low-glycemic index diet, can do better on a modified keto diet.

A health care provider should track closely; whoever is using a keto diet as a medication. A doctor and dietitian can monitor a person's progress, prescribe medications, and test for adverse effects.

The body processes fat differently from that it processes protein differently from that of carbohydrates. The Carbohydrate response to insulin is extreme. The protein response to insulin is moderate, and the fast response to the insulin is minimal. Insulin is the hormone that produces fat / conserves fat.

After you've planned out your protein and carbohydrates, eat fat. You can eat all the fat you want as long as you're not doing it to excess. But unlike Weight Watchers or other diet plans, you don't need to measure fat or count calories. Simply let your body tell you when you've had enough. If you eat fat until you're satiated, you won't have problems consuming too many calories. If you eat and still feel like you need to eat more – do it. Many beginners on keto get into trouble when they don't eat enough fat. Fasting can be incorporated into the keto diet if it's done correctly. Try out one of the intermittent fasting techniques to help accelerate and maintain weight loss after you're adapted to keto.

Electronic monitors can be beneficial to keep track of your progress at home. If you can afford it, you should get a blood sugar monitor and a ketone monitor. Track your fasting blood sugars and keep track of your ketones, ensuring they fall within the 1.5-3.0 mmol/dL range. Also, you may want to track your HDL and triglycerides. Home monitors can be used to do this and allow you to monitor progress more frequently and keep away from unnecessary trips to the doctor's office.

Lastly, remember to keep a journal. It's essential to keep track of your progress and helps you note not only how your triglycerides may be improving, but if you write down what you eat and find out you're not losing weight, it will make it easier to pinpoint problem areas where you can improve.

BREAKFAST

Bacon Cheeseburger Waffles

Preparation Time: 10 minutes

Cooking Time: 20 minutes

Servings: 4

Ingredients

Toppings

- Pepper and Salt to taste
- 12 ounces of cheddar cheese
- 4 tablespoons of sugar-free barbecue sauce
- 4 slices of bacon
- 4 ounces of ground beef, 70% lean meat and 30% fat

Waffle dough

- Pepper and salt to taste
- 3 tablespoons of parmesan cheese, grated
- 4 tablespoons of almond flour
- ¼ teaspoon of onion powder
- ¼ teaspoon of garlic powder
- 1 cup (125 g) of cauliflower crumbles
- 2 large eggs
- ounces of cheddar cheese

Directions

1. Shred about 3 ounces of cheddar cheese, then add in cauliflower crumbles in a bowl and put in half of the cheddar cheese.
2. Put into the mixture spices, almond flour, eggs, and parmesan cheese, then mix and put aside for some time.
3. Thinly cut the bacon and cook in a skillet on medium to high heat.
4. After the bacon is cooked partially, put in the beef, cook until the mixture is well done.
5. Then put the excess grease from the bacon mixture into the waffle mixture. Set aside the bacon mix.
6. Use an immersion blender to blend the waffle mix until it becomes a paste, then add into the waffle iron half of the mix and cook until it becomes crispy.
7. Repeat for the remaining waffle mixture.
8. As the waffles cook, add sugar-free barbecue sauce to the ground beef and bacon mixture in the skillet.
9. Then proceed to assemble waffles by topping them with half of the left cheddar cheese and half the beef mixture. Repeat this for the remaining waffles, broil for around 1-2 minutes until the cheese has melted then serve right away.

Nutrition:

18.8g Protein 33.9g Fats 415 Calories

Keto Breakfast Cheesecake

Preparation Time: 20 minutes

Cooking Time: 45 minutes

Servings: 24

Ingredients

Toppings

- 1/4 cup of mixed berries for each cheesecake, frozen and thawed
- Filling ingredients
- 1/2 teaspoon of vanilla extract
- 1/2 teaspoon of almond extract

- 3/4 cup of sweetener
- 6 eggs
- 8 ounces of cream cheese
- 16 ounces of cottage cheese

Crust ingredients

- 4 tablespoons of salted butter
- 2 tablespoons of sweetener

- 2 cups of almonds, whole

Directions

1. Preheat oven to around 350 degrees F.

2. Pulse almonds in a food processor then add in butter and sweetener.

3. Pulse until all the ingredients mix well and coarse dough forms.

4. Coat twelve silicone muffin pans using foil or paper liners.

5. Portion the batter evenly between the muffin pans then press into the bottom part until it forms a crust and bakes for about 8 minutes.

6. Pulse in a food processor the cream cheese and cottage cheese then pulse until the mixture is smooth.

7. Put in the extracts and sweetener then combine until well mixed.

8. Add in eggs and pulse again until it becomes smooth. Share equally the batter between the muffin pans, then bake for around 30-40 minutes.

9. Put aside until cooled completely, then put in the refrigerator for about 2 hours and then top with frozen and thawed berries.

Nutrition

12g Fats 152 Calories 6g Proteins

Egg-Crust Pizza

Preparation Time: 5 minutes

Cooking Time: 15 minutes   Servings: 2

Ingredients

- ¼ teaspoon of dried oregano to taste
- ½ teaspoon of spike seasoning to taste
- 1 ounce of mozzarella, chopped into small cubes
- 6 – 8 sliced thinly black olives
- 6 slices of turkey pepperoni, sliced into half
- 4-5 thinly sliced small grape tomatoes
- 2 eggs, beaten well
- 1-2 teaspoons of olive oil

Directions

1. Preheat the broiler in an oven than in a small bowl, beat well the eggs. Cut the pepperoni and tomatoes in slices then cut the mozzarella cheese into cubes. Drizzle oil in a skillet at medium heat, then heat the pan for around one minute until it begins to get hot. Add in eggs and season with oregano and spike seasoning, then cook for around 2 minutes.

2. Drizzle half of the mozzarella, olives, pepperoni, and tomatoes on the eggs followed by another layer of the remaining half of the above ingredients. Ensure that there is a lot of cheese on the topmost layers. Cover and cook for 4 minutes.

3. Position the pan under the preheated broiler and cook until the top has browned and the cheese has melted nicely for around 2-3 minutes. Serve immediately.

Nutrition

363 Calories 24.1g Fats 20.8g Carbohydrates

Breakfast Roll-Ups

Preparation Time: 5 minutes

Cooking Time: 15 minutes

Servings: 5

Ingredients

- Non-stick cooking spray
- 5 patties of cooked breakfast sausage
- 5 slices of cooked bacon
- cups of cheddar cheese, shredded
- Pepper and salt
- 10 large eggs

Directions

1. Prep a skillet on medium to high heat, then using a whisk, combine two of the eggs in a mixing bowl.
2. After the pan has become hot, lower the heat to medium-low heat then put in the eggs. If you want to, you can utilize some cooking spray.
3. Season eggs with some pepper and salt.
4. Seal the eggs and leave them to cook for a couple of minutes or until the eggs are almost cooked.
5. Drizzle around 1/3 cup of cheese on top of the eggs, then place a strip of bacon and divide the sausage into two and place on top.
6. Roll the egg carefully on top of the fillings. The roll-up will almost look like a taquitos. If you have a hard time folding over the egg, use a spatula to keep the egg intact until the egg has molded into a roll-up.
7. Put aside the roll-up then repeat the above steps until you have four more roll-ups; you should have 5 roll-ups in total.

Nutrition:

412.2g Calories 31.6g Fats 2.26g Carbohydrates

Basic Opie Rolls

Preparation Time: 20 minutes

Cooking Time: 35 minutes

Servings: 12

Ingredients

- 1/8 teaspoon of salt
- 1/8 teaspoon of cream of tartar
- 3 ounces of cream cheese
- 3 large eggs

Direction

1. Prepare oven to about 300 degrees, then separate the egg whites from egg yolks and place both eggs in different bowls. Using an electric mixer, beat well the egg whites until the mixture is very bubbly, then stir in the cream of tartar and mix.

2. In the bowl with the egg yolks, put in 3 ounces of cubed cheese and salt. Mix well until the mixture has doubled in size and is pale yellow. Put in the egg white mixture into the egg yolk mixture then fold the mixture gently together.

3. Spray some oil on the cookie sheet coated with some parchment paper, then add dollops of the batter and bake for around 30 minutes.

4. You will know they are ready when the upper part of the rolls is firm and golden. Put aside on a wire rack. Enjoy with some coffee.

Nutrition:

45 Calories 4g Fats 2g Proteins

Almond Coconut Egg Wraps

Preparation time: 5 minutes

Cooking time: 5 minutes

Servings: 4

Ingredients:

- 5 Organic eggs
- 1 tbsp. Coconut flour
- 25 tsp. Sea salt
- 2 tbsp. almond meal

Directions:

1. Combine the fixings in a blender and work them until creamy. Heat a skillet using the med-high temperature setting.

2. Fill two tablespoons of batter into the skillet and cook - covered about three minutes. Turn it over to cook for another 3 minutes. Serve the wraps piping hot.

Nutrition:

3g Carbohydrates

8g Protein

111 Calories

Bacon & Avocado Omelet

Preparation Time: 5 minutes

Cooking Time: 5 minutes

Servings: 1

Ingredients:

- 1 slice Crispy bacon
- 2 Large organic eggs
- 5 cup freshly grated parmesan cheese
- 2 tbsp. Ghee or coconut oil or butter
- half of 1 small Avocado

Directions:

1. Prepare the bacon to your liking and set aside. Combine the eggs, parmesan cheese, and your choice of finely chopped herbs. Warm a skillet and add the butter/ghee to melt using the medium-high heat setting. When the pan is hot, whisk and add the eggs.

2. Prepare the omelet working it towards the middle of the pan for about 30 seconds. When firm, flip, and cook it for another 30 seconds.

3. Arrange the omelet on a plate and garnish with the crunched bacon bits. Serve with sliced avocado.

Nutrition:

3.3g Carbohydrates 30g Protein 719 Calories

Bacon & Cheese Frittata

Preparation Time: 5 minutes

Cooking Time: 5 minutes

Servings: 6

Ingredients:

- 1 cup Heavy cream
- 6 Eggs
- 5 Crispy slices of bacon
- 2 Chopped green onions
- 4 oz. Cheddar cheese

Directions:

1. Warm the oven temperature to reach 350° Fahrenheit.
2. Scourge eggs and seasonings. Fill into the pie pan and top off with the remainder of the fixings. Bake 30-35 minutes. Wait for a few minutes before serving for best results.

Nutrition:

2g Carbohydrates 13g Protein 320 Calories

Bacon & Egg Breakfast Muffins

Preparation Time: 15 minutes

Cooking Time: 30 minutes

Servings: 12

Ingredients:

- 8 large Eggs
- 8 slices Bacon
- .66 cup Green onion

Directions:

1. Warm the oven at 350° Fahrenheit. Spritz the muffin tin wells using a cooking oil spray. Chop the onions and set aside.

2. Prepare a large skillet using the medium temperature setting. Fry the bacon until it's crispy and place on a layer of paper towels to drain the grease. Chop it into small pieces after it has cooled.

3. Whisk the eggs, bacon, and green onions, mixing well until all of the fixings are incorporated. Dump the egg mixture into the muffin tin (halfway full). Bake it for about 20 to 25 minutes. Cool slightly and serve.

Nutrition:

0.4g Carbohydrates 5.6g Protein 69 Calories

Bacon Hash

Preparation Time: 5 minutes

Cooking Time: 10 minutes Servings: 2

Ingredients:

- 1 Small green pepper
- 2 Jalapenos
- 1 Small onion
- 4 Eggs
- 6 Bacon slices

Directions:

1. Chop the bacon into chunks using a food processor. Set aside for now. Cut onions and peppers into thin strips. Dice the jalapenos as small as possible.

2. Heat a skillet and fry the veggies. Once browned, combine the fixings and cook until crispy. Place on a serving dish with the eggs.

Nutrition:

9g Carbohydrates 23g Protein 366 Calories

SNACK RECIPES

Sheet Pan Eggs

Preparation Time: 5 minutes

Cooking Time: 15 minutes

Servings: 4

Ingredients:

- 12 eggs
- coconut oil
- salt and pepper to taste
- ½ cup mixed bell peppers
- ¼ cup chopped chives

Directions:

1. Preheat the oven to medium (300-400°F) Grease your baking sheet with coconut oil.
2. Mix the eggs with pepper and salt in a box until it becomes frothy, and then add the bell peppers and chopped chives.
3. Transfer mixture into the pan. Bake it for 12-15 minutes. Remove it, let it cool, and then cut it into squares.

Nutrition:

13g Protein 2g Carbs 10g Fats 507 Calories

Scrambled Eggs

Preparation Time: 2 minutes

Cooking Time: 8 minutes Servings: 4

Ingredients:

- 4 oz. butter
- 8 eggs
- salt and pepper for taste

Directions:

1. Whisk together all the eggs, while seasoning it.

2. Melt the butter but not too brown. Transfer the eggs into the skillet and let it cook for 1-2 minutes, until they look and feel fluffy and creamy.

3. Tip: If you want to shake things up, you can pair this one up with bacon, salmon, or maybe avocado as well.

Nutrition:

1g Carbohydrates 31g Fat 11g Protein 327 Calories

Keto Tapas

Preparation Time: 10 minutes

Cooking Time: 5 minutes

Servings: 4

Ingredients:

- Cheese (mozzarella, cheddar, etc.)
- Cold cuts (ham, prosciutto, salami, etc.)
- Cucumber, pepper, radishes, avocado
- mayo
- 1 lemon for its juice
- pepper, salt
- Nuts (walnuts, almonds, hazelnuts)

Direction:

1. Cut your consumables into small pieces; for example, cube with the avocado as well.
2. Mix your mayo with the lemon juice, salt, and pepper. Add the toppings as you wish
3. Tip: Use up your avocado's shell, and serve this snack in it! It will look extremely classy, trust me.

Nutrition: 5g Net Carbs 1g Fiber 57g Fat 30g protein 664 Calories

Coconut Porridge

Preparation Time: 2 minutes

Cooking Time: 10 minutes

Servings: 1

Ingredients:

- 1 egg
- 1 tbsp. coconut flour
- 1 pinch ground psyllium, husk powder
- salt
- 1 oz. butter oil/coconut oil
- 4 tbsps. coconut cream

Directions:

1. Put all ingredients to a saucepan. Mix them well and place over a low heat. Stir constantly until the desired texture.

2. Serve the mixture with coconut milk or cream. You can top it with some berries.

Nutrition:

4g Carbohydrates 49g Fat 9g Protein 486 Calories

Frittata with Spinach

Preparation Time: 5 minutes

Cooking Time: 30 minutes Servings: 4

Ingredients:

- 8 eggs
- 8 oz. fresh spinach
- 5 oz. diced bacon
- 5 oz. shredded cheese
- 1 cup heavy whipping cream
- 2 tbsps. butter
- salt and pepper

Directions:

1. Preheat the oven to 350 °F. Cook bacon until crispy. Add the spinach and cook until wilted. Set them aside.

2. Mix cream and eggs, then pour it into the baking dish. Add the cheese, spinach, and bacon on the top, and place in the oven. Bake until golden brown for 25-30 minutes.

Nutrition:

4g Carbohydrates 59g Fat 27g Protein 661 Calories

Sausage Bombs

Preparation Time: 10 minutes

Cooking Time: 20 minutes

Servings: 20

Ingredients:

- 1 lb. Breakfast Sausage
- 1 Cup Almond Flour
- 1 Egg
- ¼ Cup Parmesan, Grated
- 1 Tablespoon Butter
- 2 Teaspoons Baking Powder

Directions:

1. Preheat your oven at 350 degree, and get a bowl. Mix all of your ingredients together before making twenty balls.

2. Place these sausage balls on a baking sheet, baking for twenty minutes. Serve warm or chilled.

Nutrition:

124 Calories 6g Protein 11g Fat 0.2g Net Carbs

Pesto Bombs

Preparation Time: 1 hour

Cooking Time: 1 hour and 5 minutes

Servings: 6

Ingredients:

- 1 Cup Cream Cheese, Full Fat
- 2 Tablespoons Basil Pesto
- 10 Green Olives, Sliced
- ½ Cup Parmesan Cheese, Grated

Directions:

1. Mix your butter and cream cheese. Mix all of your ingredients except parmesan cheese. Mix well, and then refrigerate for a half hour.

2. Roll in parmesan cheese before serving.

Nutrition:

123 Calories 4.3g Protein 12.9g Fat: 1.3g Net Carbs

Pork Belly Bombs

Preparation Time: 15 minutes

Cooking Time: 40 minutes

Servings: 6

Ingredients:

- ¼ Cup Mayonnaise
  - o Ounces Pork Belly, Cooked
- 3 Bacon Slices, Cut in Half
- 1 Tablespoon Horseradish, Fresh & Grated
- 1 Tablespoon Dijon Mustard
- 6 Lettuce Leaves for Serving
- Sea Salt & Black Pepper to Taste

Directions:

1. Preheat your oven to 325, and then cook your bacon for a half hour. Allow it to cool, and then crumble your bacon. Place it in a dish.

2. Shred your pork belly, placing it in a bowl. Add in your mayonnaise, horseradish, and mustard. Mix and seasoned it.

3. Divide this mixture into six mounds, and then roll it in your crumbled bacon.

4. Serve on lettuce leaves.

Nutrition:

263 Calories 3.5g Protein 26.4g Fat 0.3g Net Carbs

Cheesy Artichoke Bombs

Preparation Time: 20 minutes

Cooking Time: 50 minutes

Servings: 4

Ingredients:

- 2 Bacon Slices
- 2 Tablespoons Ghee
- 1 Clove Garlic, Minced
- ½ Onion, Large, Peeled & Diced
- 1/3 Cup Artichoke Hearts, Canned & Sliced
- ¼ Cup Sour Cream
- 1 Tablespoon Lemon Juice, Fresh
- ¼ Cup Mayonnaise
- ¼ Cup Swiss Cheese, Grated
- 4 Avocado Halves, Pitted
- Sea Salt & Black Pepper to Taste

Directions:

1. Fry your bacon for five minutes. It should be crisp. Allow it to cool before crumbling it and placing it in a bowl. Set the bowl to the side.

2. Cook your garlic and onion in the same pan using your ghee for three minutes. Combine this in with your bacon, and then throw in your remaining ingredients.

3. Mix well, seasoning with salt and pepper. Refrigerate your mixture for a half hour before filling each avocado half with one.

Nutrition:

408 Calories 6.8g Protein 39.6g Fat 4g Net Carbs

Sausage and Avocado Bombs

Preparation Time: 20 minutes

Cooking Time: 55 minutes

Servings: 4

Ingredients:

- 12 Ounces Spanish Chorizo Sausage, Diced
- ¼ Cup Butter, Unsalted
- 2 Hardboiled Eggs, Large & Diced
- 2 Tablespoons Mayonnaise
- 2 Tablespoons Chives, Chopped
- 1 Tablespoon Lemon Juice, Fresh
- 4 Avocado Halves, Pitted
- Sea Salt to Taste
- Cayenne Pepper to Taste

Directions:

1. Fry your chorizo over heat for five minutes before placing it to the side. Get out a bowl and combine all of your ingredients, mashing it together. Make sure not to add in your avocado halves. They're for serving.

2. Refrigerate this mixture for a half hour before filling each avocado half. Serve chilled.

Nutrition:

419 Calories 11.4g Protein 38.9g Fat 2.7g Net Carbs

LUNCH RECIPES

Cheesy Ham Quiche

Preparation Time: 10 minutes

Cooking Time: 30 minutes

Servings: 6

Ingredients:

- 8 Eggs
- 1 cup Zucchini
- ½ cup heavy Cream
- 1 cup Ham
- 1 tsp. Mustard

Directions:

1. Prep stove to 375 and get pie plate for your quiche. Shred the zucchini.
2. Once done, drain. Place the zucchini into your pie plate along with the cooked ham pieces and cheese. Whisk the seasonings, cream, and eggs. Pour on top then cook for 40 minutes.
3. If the quiche is cooked to your liking, take the dish from the oven and allow it to chill slightly before slicing.

Nutrition:

25g Fats 2g Carbohydrates 20g Proteins

Loaded Cauliflower Rice

Preparation Time: 10 minutes

Cooking Time: 20 minutes

Servings: 4

Ingredients:

- 1 Cauliflower
- 1 cup Cheddar Cheese
- 1 lb. Bacon
- ½ cup Chives

Directions:

1. Rice your cauliflower. You can choose to do this by hand. Cook the bacon in a grilling pan over a medium heat.

2. Place your cauliflower rice into a microwave-safe bowl and sprinkle your shredded cheese over the top.

3. Place bowl into the microwave for a minute and allow for the rice to cook through and the cheese to melt.

4. Top with your bacon pieces and season to your liking.

Nutrition:

10g Fats 5g Carbohydrates 5g Proteins

Creamy Garlic Chicken

Preparation Time: 5 minutes

Cooking Time: 15 minutes Servings: 4

Ingredients:

- 4 chicken breasts
- 1 tsp. garlic powder
- 1 tsp. paprika
- 2 tbsp. butter
- 1 tsp. salt
- 1 cup heavy cream
- ½ cup sun-dried tomatoes
- 2 cloves garlic
- 1 cup spinach

Directions:

1. Blend the paprika, garlic powder, and salt and rub both sides of the chicken.
2. Melt the butter in a frying pan over medium heat. Fry chicken for 5 minutes each side. Set aside.
3. Whisk the heavy cream, sun-dried tomatoes, and garlic. Cook for 2 minutes. Sauté spinach for additional 3 minutes. Place chicken back to the pan and cover with the sauce.

Nutrition:

12g Carbohydrates 26g Fat 4g Protein

Chinese Pork Bowl

Preparation Time: 5 minutes

Cooking Time: 15 minutes

Servings: 4

Ingredients:

- 1 ¼ pounds pork belly
- 2 Tbsp. tamari soy sauce
- 1 Tbsp. rice vinegar
- 2 cloves garlic
- 3 oz. butter
- 1-pound Brussels sprouts
- ½ leek

Directions:

1. Fry the pork over medium-high heat.
2. Combine the garlic cloves, butter, and Brussels sprouts. Add in to the pan and cook.
3. Drizzle soy sauce and rice vinegar together and pour into the pan. Season.
4. Top with chopped leek.

Nutrition:

7g Carbohydrates 97g Protein 993 Calories

Relatively Flavored Gratin

Preparation Time: 15 minutes

Cooking Time: 46 minutes Servings: 8

Ingredients:

- ½ C. heavy whipping cream
- 2 tbsp. butter
- ½ tsp. garlic powder
- ¼ tsp. xanthan gum
- 4 C. zucchini
- 1 small yellow onion
- 1½ C. pepper jack cheese

Directions:

1. Prepare oven to 375 0 F and grease a 9×9-inch baking dish. In a microwave-safe dish, mix heavy whipping cream, butter, garlic powder, and xanthan gum and melt 1 minute.

2. Arrange 1/3 of zucchini and onion slices at the bottom and season and ½ C. of pepper jack cheese. Repeat the layers twice. Spread the cream mixture on top evenly. Bake for 45 minutes. Remove the baking dish from oven and set aside.

Nutrition:

140 Calories 3.9g Carbohydrates 5.5g Protein

Low Carb Crack Slaw Egg Roll

Preparation time: 10 minutes

Cooking Time: 20 minutes

Serving: 2

Ingredients:

- 1 lb. ground beef
- 4 cups shredded coleslaw mix
- 1 tbsp. avocado oil
- 1 tsp. sea salt
- ¼ tsp. black pepper
- 4 cloves garlic, minced
- 3 tbsp. fresh ginger, grated
- ¼ cup coconut amines
- 2 tsp. toasted sesame oil
- ¼ cup green onions

Directions:

1. Cook avocado oil over medium-high heat. Cook the garlic.
2. Cook ground beef for 10 minutes. Season well.
3. Once cooked, you can lower the heat and add the coleslaw mix and the coconut amines. Stir for 5 minutes.
4. Garnish green onions and the toasted sesame oil.

Nutrition:

104 calories 5g fat 18g protein

Low Carb Beef Stir Fry

Preparation Time: 10 minutes

Cooking Time: 25 minutes

Serving: 3

Ingredients:

- ½ cup zucchini
- ¼ cup organic broccoli florets
- 1 bunch baby bok choy
- 2 tbsp. avocado oil
- 2 tsp. coconut amines
- 1 small ginger
- 8 oz. skirt steak

Directions:

1. Heat the pan and add 1 tbsp. oil. Sear steak on high heat for 2 minutes per side.
2. Set to medium heat and cook the broccoli, ginger, ghee, and coconut amines.
3. Cook the book choy for another minute
4. Mix in zucchini and cook.

Nutrition:

104 calories 6g fat 31g protein

Pesto Chicken and Veggies

Preparation Time: 10 minutes

Cooking Time: 35 minutes

Serving: 3

Ingredients:

- 2 tbsp. olive oil
- 1 cup cherry tomatoes
- ¼ cup basil pesto
- 1/3 cup sun-dried tomatoes
- 1-pound chicken thigh
- 1-pound asparagus

Directions:

1. Preheat two tablespoons of olive oil and sliced chicken on medium heat. Season and add ½ cup of the sun-dried tomatoes.
2. Cook well. Spoon out the chicken and tomatoes and put them in a separate container.
3. Place the asparagus in same skillet and pour in the pesto. Put heat on medium and add the remaining sun-dried tomatoes. Cook for 10 minutes. Put it on a separate plate.
4. Position the chicken back in the pan and pour in pesto. Stir over medium heat for 2 minutes.

Nutrition:

104 calories 8g fat 26g protein

Crispy Peanut Tofu and Cauliflower Rice Stir-Fry

Preparation Time: 10 minutes

Cooking Time: 80 minutes

Serving: 4

Ingredients:

- 12 oz. tofu
- 1 tbsp. sesame oil

- 2 cloves garlic
- 1 small cauliflower head

For the sauce:

- 1 ½ tbsp. sesame oil
- ½ tsp. chili garlic sauce
- 2 ½ tbsp. peanut butter

- ¼ cup low sodium soy sauce
- ½ cup light brown sugar

Directions:

1. Strain tofu for 90 minutes.
2. Preheat the oven to 400 degrees Fahrenheit. Cube the tofu, and prepare your baking sheet.
3. Bake for 25 minutes and allow it to cool.
4. Combine the sauce ingredients.
5. Put the tofu in the sauce and coat the tofu thoroughly. Leave it for 15 minutes.
6. Shred the cauliflower into rice- size bits.
7. Situate skillet on medium heat. Cook veggies on a bit of sesame oil and soy sauce. Set aside.
8. Put tofu on the pan. Stir frequently. Set aside.
9. Steam your cauliflower rice for 8 minutes. Stir some sauce.
10. Mix ingredients together. Mix cauliflower rice with the veggies and tofu. Serve.

Nutrition:

107 calories 9g fat 30g protein

Keto Fried Chicken

Preparation Time: 10 minutes

Cooking Time: 45 minutes Serving: 4

Ingredients:

- 4 chicken thighs
- Frying oil
- 2 large eggs
- 2 tbsp. heavy whipping cream

For the breading:

- 2/3 cup parmesan cheese
- 2/3 cup almond flour
- 1 tsp. salt
- ½ tsp. black pepper
- ½ tsp. cayenne
- ½ tsp. paprika

Directions:

1. Beat together the eggs and heavy cream.
2. Mix all the breading ingredients. Set aside.
3. Cut the chicken thigh into 3 even pieces and pat dry with paper towel.
4. Dip the chicken in the bread first before dipping it in the egg wash and then finally, dipping it in the breading again.
5. Fill 2 inches of oil in a pot and preheat at 350 degrees Fahrenheit. Gradually lower the heat.
6. Put the coated chicken in your hot oil. Fry for 5 minutes.
7. Strain cooked chicken.
8. Try not to overcrowd the pan. Serve.

Nutrition:

104 calories 5g fat 29g protein

SALAD RECIPES

Smoked Salmon Salad

Preparation Time: 17 minutes

Cooking Time: 10 minutes Servings: 4

Ingredients:

- 8 oz. smoked salmon, sliced into thin pieces
- 2 oz. pecans, crushed
- 3 medium tomatoes, chopped
- ½ cup lettuce, chopped
- 1 cucumber, diced
- 1/3 cup cream cheese
- 1/3 cup coconut milk
- ½ tsp. oregano
- 1 tbsp. lemon juice, chopped
- ½ tsp. basil
- 1 tsp. salt

Directions:

1. In medium bowl, combine salmon with pecans and stir. Add tomatoes, lettuce and cucumber, stir well.
2. In another bowl, mix together cream cheese, coconut milk, oregano, lemon juice, basil and salt. Stir mixture until get homogenous mass. Serve salmon salad with cream cheese sauce.

Nutrition:

211 Calories 15.9g Fat 9.95g Protein

Tuna Salad

Preparation Time: 18 minutes

Cooking Time: 10 minutes

Servings: 4

Ingredients:

- 1 can tuna
- 4 eggs, boiled, peeled and chopped
- 1 oz. olives, pitted and sliced
- 1/3 cup cheese cream
- ½ cup almond milk
- ½ tsp. ground black pepper
- ½ tsp. kosher salt
- 1 tbsp. garlic, minced

Directions:

1. In medium bowl, mash tuna with fork. Add chopped eggs and stir. Add sliced olives and stir.

2. In another bowl, whisk together cheese cream and almond milk. Add black pepper, salt and garlic, stir carefully. Add cheese mixture to tuna mixture and mix up. Serve.

Nutrition:

182 Calories 11.9g Fat 12.88g Protein

Caprese Salad

Preparation Time: 7 minutes

Cooking Time: 10 minutes

Servings: 2

Ingredients:

- 8 oz. mozzarella cheese
- 1 medium tomato
- 4 basil leaves
- Salt and ground black pepper to taste
- 3 tsp. balsamic vinegar
- 1 tbsp. olive oil

Directions:

1. Slice mozzarella cheese and tomato. Torn basil leaves. Alternate tomato and mozzarella slices on 2 plates. Season with pepper and salt. Drizzle vinegar and olive oil. Sprinkle with the basil leaves.
2. Serve.

Nutrition:

148 Calories 5.9g Carbs 11.8g Fat 8.95g Protein

Warm Bacon Salad

Preparation Time: 16 minutes

Cooking Time: 18 minutes

Servings: 5

Ingredients:

- 16 oz. bacon strips, chopped
- 1 tsp. cilantro
- 1 tsp. ground ginger
- 1 tsp. kosher salt
- 2 tbsp. butter
- 3 boiled eggs, peeled and chopped
- 2 tomatoes, diced
- 1 oz. spinach, chopped
- 4 oz. Cheddar cheese, grated
- 1 tsp. almond milk
- 7 oz. eggplant, peeled and diced

Directions:

1. In medium bowl, combine bacon, cilantro, ginger and salt. Heat up pan over medium heat and melt 1 tablespoon of butter.
2. Put bacon in pan and cook for 5 minutes. Transfer bacon to plate.
3. Meanwhile, in bowl, mix together chopped eggs, tomatoes and spinach. Sprinkle with cheese and add almond milk. Heat up pan again over medium heat and melt remaining 1 tablespoon of butter.
4. Add diced eggplants and fry for 8 minutes, stirring occasionally. Then add bacon and roasted eggplants to salad. Season with salt and stir gently. Serve.

Nutrition:

159 Calories 4.22g Carbs 13g Fat 8.75g Protein

Cauliflower Side Salad

Preparation Time: 14 minutes

Cooking Time: 7 minutes Servings: 8

Ingredients:

- 21 oz. cauliflower
- 1 tbsp. water
- 4 boiled eggs, peeled and chopped
- 1 cup onion, chopped
- 1 cup celery, chopped
- 1 cup mayonnaise
- Salt and ground black pepper to taste
- 2 tbsp. cider vinegar
- 1 tsp. sucralose

Directions:

1. Divide cauliflower into florets and put them in heatproof bowl. Add water and place in microwave, cook for 5 minutes. Transfer to serving bowl.
2. Add eggs, onions, and celery. Stir gently.

3. In another bowl, whisk together mayonnaise, black pepper, salt, vinegar and sucralose. Add this sauce to salad and toss to coat. Serve.

Nutrition:

209 Calories 2.9g Carbs 19.7g Fat 3.97g Protein

Keto Tricolor Salad

Preparation Time: 12 minutes

Cooking Time: 8 minutes

Servings: 5

Ingredients:

- 5 oz. mozzarella cheese
- 1 tsp. oregano
- 1 tsp. minced garlic
- 1 tsp. basil
- 1 tbsp. coconut oil
- 1 tsp. lemon juice
- 2 medium tomatoes, sliced
- 7 oz. avocado, pitted and sliced
- 8 olives, pitted and sliced

Directions:

1. Cut mozzarella cheese balls into halves. In medium bowl, mix together oregano, garlic, basil, coconut oil and lemon juice.

2. On serving plate place sliced tomato, then place sliced avocado and olives. Put mozzarella pieces on top. Drizzle coconut sauce over salad and serve.

Nutrition:

239 Calories 7.9g Carbs 20.1g Fat 11.77g Protein

Low Carb Chicken Salad with Chimichurri Sauce

Preparation Time: 10 minutes

Cooking Time: 15 minutes

Serving: 5

Ingredients:

- 250 grams of various lettuce leaves
- 2 medium chicken breasts
- 2 medium avocados
- ¼ cup olive oil
- 3 tablespoons red wine vinegar
- ¼ cup parsley
- 1 tablespoon oregano
- 1 teaspoon chili pepper
- 1 teaspoon garlic

Directions:

1. Preheat the non-stick skillet. Place the lettuce and diced avocado in a salad bowl. Cook the chicken breasts, fry them until white. Let the chicken cool.
2. In a small bowl, combine olive oil, vinegar, parsley, oregano, garlic, and chili. Cut the chicken breast into cubes. Add the chopped chicken fillet to the salad and season with the classic chimichurri sauce.
3. Garnish the salad with additional chimichurri sauce and serve.

Nutrition:

285.94 calories 21.24 g fat 17.24 g protein.

Potluck Lamb Salad

Preparation Time: 20 minutes

Cooking Time: 10 minutes

Servings: 4

Ingredients:

- 2 tbsp. olive oil, divided
- 12 oz. grass-fed lamb leg steaks
- 6½ oz. halloumi cheese
- 2 jarred roasted red bell peppers
- 2 cucumbers, cut into thin ribbons
- 3 C. fresh baby spinach
- 2 tbsp. balsamic vinegar

Directions:

1. In a skillet, heat 1 tbsp. of the oil over medium-high heat and cook the lamb steaks for about 4-5 minutes per side or until desired doneness. Transfer the lamb steaks onto a cutting board for about 5 minutes. Then cut the lamb steaks into thin slices. In the same skillet, add haloumi and cook for about 1-2 minutes per side or until golden.

2. In a salad bowl, add the lamb, haloumi, bell pepper, cucumber, salad leaves, vinegar, and remaining oil and toss to combine.

3. Serve immediately.

Nutrition:

420 Calories

35.4g Protein

1.3g Fiber

Spring Supper Salad

Preparation Time: 15 minutes

Cooking Time: 5 minutes

Servings: 5

Ingredients:

For Salad:

- 1 lb. fresh asparagus
- ½ lb. smoked salmon
- 2 heads red leaf lettuce
- ¼ C. pecans

For Dressing:

- ¼ C. olive oil
- 2 tbsp. fresh lemon juice
- 1 tsp. Dijon mustard

Directions:

1. In a pan of boiling water, add the asparagus and cook for about 5 minutes. Drain the asparagus well. In a serving bowl, add the asparagus and remaining salad ingredients and mix. In another bowl, add all the dressing ingredients and beat until well combined. Place the dressing over salad and gently, toss to coat well. Serve immediately.

Nutrition:

223 Calories 8.5g Carbohydrates 3.5g Fiber

Chicken-of-Sea Salad

Preparation Time: 15 minutes

Cooking Time: 5 minutes

Servings: 6

Ingredients:

- 2 (6-oz.) cans olive oil-packed tuna
- 2 (6-oz.) cans water packed tuna
- ¾ C. mayonnaise
- 2 celery stalks
- ¼ of onion
- 1 tbsp. fresh lime juice
- 2 tbsp. mustard
- 6 C. fresh baby arugula

Directions:

1. In a large bowl, add all the ingredients except arugula and gently stir to combine. Divide arugula onto serving plates and top with tuna mixture. Serve immediately.

Nutrition:

325 Calories 27.4g Protein 1.1g Fiber

DINNER RECIPES

Parmesan-Crusted Halibut with Asparagus

Preparation Time: 10 minutes

Cooking Time: 15 minutes

Servings: 4

Ingredients:

- 2 tablespoons olive oil
- ¼ cup butter, softened
- Salt and pepper
- ¼ cup grated Parmesan
- 1-pound asparagus, trimmed
- 2 tablespoons almond flour
- 4 (6-ounce) boneless halibut fillets
- 1 teaspoon garlic powder

Directions:

1. Preheat the oven to 400 F and line a foil-based baking sheet. Throw the asparagus in olive oil and scatter over the baking sheet.

2. In a blender, add the butter, Parmesan cheese, almond flour, garlic powder, salt and pepper, and mix until smooth. Place the fillets with the asparagus on the baking sheet, and spoon the Parmesan over the eggs.

3. Bake for 10 to 12 minutes, then broil until browned for 2 to 3 minutes.

Nutrition:

415 Calories 26g Fats 42g Protein 3g Carbohydrates

Hearty Beef and Bacon Casserole

Preparation Time: 25 minutes

Cooking Time: 30 minutes

Servings: 8

Ingredients:

- 8 slices uncooked bacon
- 1 medium head cauliflower, chopped
- ¼ cup canned coconut milk
- Salt and pepper
- 2 pounds ground beef (80% lean)
- 8 ounces mushrooms, sliced
- 1 large yellow onion, chopped
- 2 cloves garlic, minced

Directions:

1. Preheat to 375 F on the oven. Cook the bacon in a skillet until crispy, then drain and chop on paper towels.

2. Bring to boil a pot of salted water, then add the cauliflower. Boil until tender for 6 to 8 minutes then drain and add the coconut milk to a food processor. Mix until smooth, then sprinkle with salt and pepper.

3. Cook the beef until browned in a pan, then wash the fat away. Remove the mushrooms, onion, and garlic, then move to a baking platter. Place on top of the cauliflower mixture and bake for 30 minutes. Broil for 5 minutes on high heat, then sprinkle with bacon to serve.

Nutrition:

410 Calories 25g Fats 37g Protein 6g Carbohydrates

Sesame Wings with Cauliflower

Preparation Time: 5 minutes

Cooking Time: 30 minutes

Servings: 4

Ingredients:

- 2 ½ tablespoons soy sauce
- 2 tablespoons sesame oil
- 1 ½ teaspoons balsamic vinegar
- 1 teaspoon minced garlic
- 1 teaspoon grated ginger
- Salt
- 1-pound chicken wing, the wings itself
- 2 cups cauliflower florets

Directions:

1. In a freezer bag, mix the soy sauce, sesame oil, balsamic vinegar, garlic, ginger, and salt, then add the chicken wings. Coat flip, then chill for 2 to 3 hours.

2. Preheat the oven to 400 F and line a foil-based baking sheet. Spread the wings along with the cauliflower onto the baking sheet. Bake for 35 minutes, then sprinkle on to serve with sesame seeds.

Nutrition:

400 Calories 15g Fats 5g Protein 3g Carbohydrates

Fried Coconut Shrimp with Asparagus

Preparation Time: 15 minutes

Cooking Time: 10 minutes   Servings: 6

Ingredients:

- 1 ½ cups shredded unsweetened coconut - 2 large eggs
- Salt and pepper
- 1 ½ pounds large shrimp, peeled and deveined
- ½ cup canned coconut milk
- 1-pound asparagus, cut into 2-inch pieces

Directions:

1. Pour the coconut onto a shallow platter. Beat the eggs in a bowl with a little salt and pepper. Dip the shrimp into the egg first, then dredge with coconut.

2. Heat up coconut oil over medium-high heat in a large skillet. Add the shrimp and fry over each side for 1 to 2 minutes until browned.

3. Remove the paper towels from the shrimp and heat the skillet again. Remove the asparagus and sauté to tender-crisp with salt and pepper, then serve with the shrimp.

Nutrition:

535 Calories 38g Fats 16g Protein 3g Carbohydrates

Coconut Chicken Curry with Cauliflower Rice

Preparation time: 15 minutes

Cooking time: 30 minutes

Servings: 6

Ingredients:

- 1 tablespoon olive oil
- 1 medium yellow onion, chopped
- 1 ½ pounds boneless chicken thighs, chopped
- Salt and pepper
- 1 (14-ounce) can coconut milk
- 1 tablespoon curry powder
- 1 ¼ teaspoon ground turmeric
- 3 cups riced cauliflower

Directions:

1. Heat the oil over medium heat, in a large skillet. Add the onions, and cook for about 5 minutes, until translucent.

2. Stir in the chicken and season with salt and pepper-cook for 6 to 8 minutes, stirring frequently until all sides are browned. Pour the coconut milk into the pan, then whisk in the curry and turmeric powder.

3. Simmer until hot and bubbling, for 15 to 20 minutes. Meanwhile, steam the cauliflower rice until tender with a few tablespoons of water. Serve the cauliflower rice over the curry.

Nutrition:

430 Calories 29g Fats 9g Protein 3g Carbohydrates

Grilled Whole Chicken

Preparation Time: 20 minutes

Cooking Time: 20 minutes

Servings: 6

Ingredients

- ¼ cup butter

- 2 tablespoons lemon juice

- 2 teaspoons fresh lemon zest

- 1 teaspoon dried oregano

- 2 teaspoons paprika

- 1 teaspoon onion powder

- 1 teaspoon garlic powder

- Salt and ground black pepper

- 1 (4-pound) grass-fed whole chicken

Directions:

1. Preheat the grill to medium heat. Grease the grill grate. Place onto a large cutting board, breast-side down. Mix butter, lemon juice, lemon zest, oregano, spices, salt, and black pepper.

2. Cut both sides of backbone. Remove the backbone. Flip and open. Firmly press breast to flatten.

3. Coat the whole chicken with the oil mixture. Grill for 20 minutes.

4. Remove from the grill and set aside for 10 minutes.

Nutrition:

532 Calories 17g Fat 0.5g Fiber

Grilled Chicken Breast

Preparation Time: 15 minutes

Cooking Time: 14 minutes

Servings: 4

Ingredients

- ¼ cup balsamic vinegar
- 2 tablespoons olive oil
- 1½ teaspoons lemon juice
- ½ teaspoon lemon-pepper seasoning
- 4 (6-ounce) grass-fed chicken breast halves

Directions:

1. Blend vinegar, oil, lemon juice, and seasoning. Coat chicken breasts with the mixture. Marinate for 30 minutes.

2. Preheat and grease the grill to medium heat. Place the chicken breasts onto the grill and cover.

3. Cook for 7 minutes. Serve.

Nutrition:

258 Calories 11.3g Fat 0.1g Fiber

Glazed Chicken Thighs

Preparation Time: 15 minutes

Cooking Time: 35 minutes

Servings: 8

Ingredients

- ½ cup balsamic vinegar

- 1/3 cup low-sodium soy sauce

- 3 tablespoons Yukon syrup

- 4 tablespoons olive oil

- 3 tablespoons chili sauce

- 2 tablespoons garlic

- Salt and black pepper

- 8 (6-ounce) grass-fed chicken thighs

Directions:

1. Mix all ingredients (except chicken thighs and sesame seeds). Mix half of marinade and chicken thighs. Seal and shake well.

2. Chill for 1 hour. Chill remaining marinade. Preheat oven to 425°F.

3. Mix reserved marinade over medium heat and boil. Cook for 5 minutes. Remove and set aside.

4. Remove from the bag and discard excess marinade. Arrange chicken thighs into a 9x13-inch baking dish in a single layer and coat with cooked marinade.

5. Bake for 30 minutes. Serve.

Nutrition:

406 Calories 19.6g Fat 0.1g Fiber

Bacon-Wrapped Chicken Breasts

Preparation Time: 15 minutes

Cooking Time: 33 minutes Servings: 4

Ingredients

Chicken Marinade

- 3 tablespoons balsamic vinegar
- 3 tablespoons olive oil
- 2 tablespoons water
- 1 garlic clove

- 1 teaspoon dried Italian seasoning
- ½ teaspoon dried rosemary
- 4 (6-ounce) grass-fed chicken breasts

Stuffing

- 16 fresh basil leaves
- 1 large fresh tomato
- 4 provolone cheese slices

- 8 bacon slices
- ¼ cup Parmesan cheese

Directions:

1. For marinade:

    a. Mix all ingredients (except chicken).

2. For chicken

3. Chop chicken breast horizontally, without cutting all the way through.

4. Repeat with the remaining chicken breasts. Coat with marinade. Chill 30 minutes.

5. Preheat your oven to 500°F. Grease baking dish.

6. Place 4 basil leaves onto the bottom half of a chicken breast. Followed by 3 tomato slices and 1 provolone cheese slice. Fold the top half over filling.

7. Wrap the breast with 3 bacon slices. Repeat. Situate into the prepared baking dish in a single layer.

8. Bake for 30 minutes. Remove and sprinkle with Parmesan cheese evenly. Bake for 3 minutes more.

Nutrition:

633 Calories 36g Fat 0.3g Fiber

Chicken Parmigiana

Preparation Time: 15 minutes

Cooking Time: 26 minutes

Servings: 5

Ingredients

- 5 (6-ounce) grass-fed chicken breasts
- 1 large organic egg
- ½ cup superfine almond flour
- ¼ cup Parmesan cheese,
- ½ teaspoon dried parsley
- ½ teaspoon paprika
- ½ teaspoon garlic powder
- 1 cup sugar-free tomato sauce
- 5 ounces mozzarella cheese
- 2 tablespoons fresh parsley

Directions:

1. Preheat your oven to 375ºF. Wrap 1 chicken breast in parchment paper. Pound the chicken breast into ½-inch thickness

2. Repeat with the rest. Put the beaten egg. Mix almond flour, Parmesan, parsley, spices, salt, and black pepper in another dish.

3. Dip it into the whipped egg and coat with the flour mixture. Heat the oil over medium-high heat and fry for 3 minutes.

4. Dry chicken breasts. Spread at bottom of a casserole dish ½ cup of tomato sauce. Arrange the chicken breasts over marinara sauce in a single layer.

5. Drizzle with the remaining tomato sauce, mozzarella cheese slices. Bake for 20 minutes. Garnish with parsley. Serve

Nutrition:

458 Calories 25.4g Fat 7.9g Carbs

Conclusion

I f you're eyeing for a diet that works and gives you the results you want, then it's time to take your health and performance to the next level. It is also one of the most effective ways to reduce appetite and feel full. It's also a natural heal for diabetes, epilepsy, and Alzheimer's disease.

This keto diet is a low-carbohydrate, high-fat diet that increases your body's ability to burn fat as fuel.

The ketogenic diet main purpose is to cause your body to make ketones, which are compounds produced by the liver used as an alternative fuel source for your body instead of glucose (sugar). These ketones then serve as a fuel source all over the body, especially for the brain.

In less than 5 years, the keto diet has gone from a notorious fad diet to a well-respected high-protein health and wellness regimen. An increasing amount of people are deciding on living without carbohydrates. They rely solely on fat-forming foods like meat, fish, eggs, cheese, butter, and coconut oil for their caloric intake. This trend has been gaining ground for more than 30 years with people following such diets as Atkins. The reason for this recent spike in popularity can be attributed partially to the 2014 documentary The Carbohydrate Addict, which focuses on Dr. Robert Atkins' theory that carbohydrates play a central role in heart disease.

Not long after its release, the ketogenic diet was used as the backbone of a new trend known as "ketogenic" or "low carb" diets. These low-carb diets claim that by restricting carbohydrates from your daily meal plan, your body will become efficient at burning fat as fuel instead of glucose. The purpose of following such a regimen is to create your own "ketogenic" state wherein your body will naturally become efficient at burning fat stored within your liver and muscles for energy instead of carbohydrates. The first thing people should be prepared for is the signs of the body-switching over to ketosis. These include bad breath, weight loss, appetite decrease, and potential weakness in the beginning stages. It is normal to have these reactions while doing the keto diet. It can also be helpful to be familiar with the keto flu's signs and symptoms, which can affect people at varying severities. Finally, they should have an idea of how long they will need to stay on a diet to achieve their desired results. Some people choose to do standard keto until they reach their weight loss goals and then choose a less vigorous form of the diet to keep the pounds off.

For people who have stomach issues when starting the diet, switching to fats that are easier to digest can be a smart move for the beginning stages. Adding fiber to the diet can also help regulate the gut and ease those uncomfortable symptoms.

After making it through the keto flu, here are the benefits of the diet. Some people decide to stay on the meal plan long term. Although it is not recommended to do full keto for longer than a year, keeping some form of the diet long-term can help to ensure the goals met are not lost. To ensure that staying on a diet is simple and easy, people should focus on eating quality fats that smoothly help their brain and body function. If the body doesn't have to work hard to digest food, the person will usually have more energy and feel better overall.

CPSIA information can be obtained
at www.ICGtesting.com
Printed in the USA
LVHW011347250621
691051LV00016B/1355